PETS IN THE WILD!

LEOPARD GECKOS IN THE WILD!

by Grace Hansen

Cody Koala
An Imprint of Pop!
popbooksonline.com

Hello! My name is Cody Koala

This book is filled with videos, puzzles, games, and more! Scan the QR codes* while you read, or visit the website below to make this book pop.

popbooksonline.com/l-gecko

*Scanning QR codes requires a web-enabled smart device with a QR code reader app and a camera.

abdobooks.com

Published by Pop!, a division of ABDO, PO Box 398166, Minneapolis, Minnesota 55439.

Printed in the United States of America, North Mankato, Minnesota.

052024
092024

THIS BOOK CONTAINS RECYCLED MATERIALS

Cover Photo: Shutterstock Images
Interior Photos: Shutterstock Images, Getty Images
Editor: Elizabeth Andrews
Series Designer: Laura Graphenteen; Neil Klinepier

Library of Congress Control Number: 2023947484

Publisher's Cataloging-in-Publication Data

Names: Hansen, Grace, author.
Title: Leopard Geckos in the wild! / by Grace Hansen
Description: Minneapolis, Minnesota : Pop!, 2025 | Series: Pets in the wild! | Includes online resources and index
Identifiers: ISBN 9781098246150 (lib. bdg.) | ISBN 9781098246716 (ebook)
Subjects: LCSH: Leopard gecko--Juvenile literature. | Wild animals--Juvenile literature. | Wild animals as pets--Juvenile literature. | Geckos--Behavior--Juvenile literature. | Lizards--Behavior--Juvenile literature. | Reptiles--Juvenile literature.
Classification: DDC 636.0887--dc23

Table of Contents

Chapter 1
Leopard Geckos 4

Chapter 2
Wild Leopard Geckos 8

Chapter 3
Pet Leopard Geckos 14

Chapter 4
Caring for
Leopard Geckos 16

Making Connections 22
Glossary 23
Index 24
Online Resources 24

Chapter 1

Leopard Geckos

Leopard geckos are ground-dwelling lizards. Their bodies are often yellowish in color and covered in black spots. This is how they got their name!

The underside of a leopard gecko's body is white in color.

Watch a video here!

Leopard geckos are **carnivores**. They eat insects, such as crickets and grasshoppers. They also eat smaller reptiles and young mice.

Chapter 2

Wild Leopard Geckos

Leopard geckos are originally from the Middle East and Asia. Today, they are most often found in Afghanistan, Iran, Pakistan, and India. They live in warm and dry areas.

Where Wild Leopard Geckos Live

Learn more here!

Leopard geckos have very good senses of hearing and sight. This keeps them safe from predators.

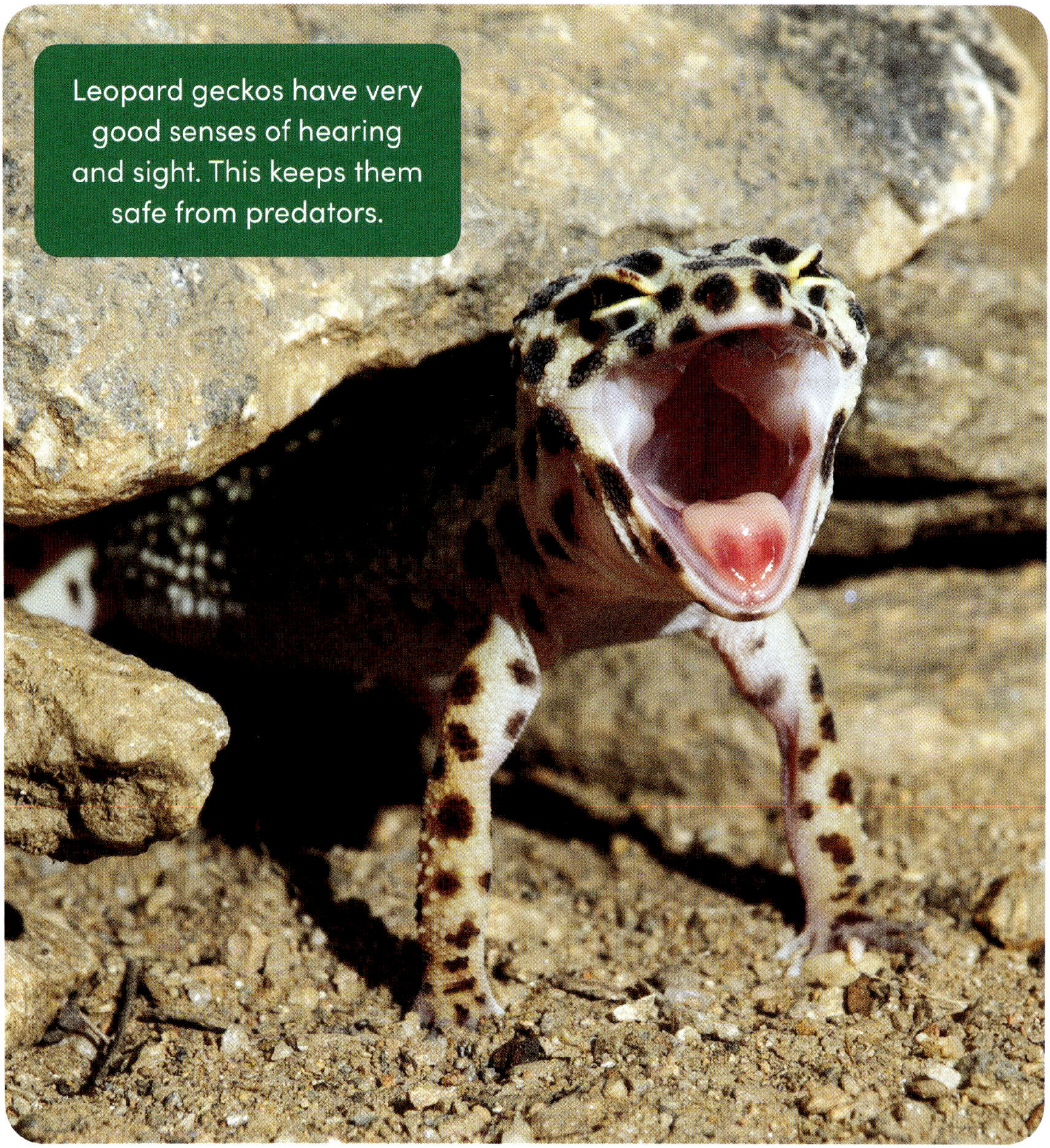

Leopard geckos spend most of their days hidden from **predators** and the sun. They come out in the early morning or evening hours. This is when the temperature is just right for them.

Leopard geckos have special, thick tails. Their tails store fat. This helps keep them healthy when there is not enough food for them to eat.

Chapter 3

Pet Leopard Geckos

Leopard geckos are one of the most common pet lizards. They are known for being easy-going and hardy. They have been **bred** as pets for more than 50 years.

Explore links here!

Leopard geckos can live for up to 20 years.

Chapter 4

Caring for Leopard Geckos

Leopard geckos are easy to care for. Their **enclosures** should mimic their natural **environment**. Half of the enclosure should be kept warm using special heaters or bulbs.

The warmer half of the enclosure should be kept between 88–90 °F (31.1–32.2 °C).

Complete an activity here!

The floor of a leopard gecko's enclosure should be mostly open. This allows the lizard to move around freely. It should also be easy to clean. Common flooring includes paper towels and special reptile carpet or tile.

Leopard geckos should have places to hide. Their water dish should be cleaned and refilled daily.

Leopard geckos should be fed mostly live and moving insects and larvae, such as crickets and mealworms.

Making Connections

Text-to-Self

What do you like best about leopard geckos?

Text-to-Text

Have you read any other books about leopard geckos? What did you learn in those books that was not in this one?

Text-to-World

There are many cool reptiles that people keep as pets, including all kinds of snakes and lizards. What is your favorite pet reptile?

Glossary

bred – caused an animal to be born.

carnivore – an animal that only eats meat.

enclosure – a type of secure container meant to keep a pet animal safe.

environment – the natural world, including air, water, land, and animals.

predator – an animal that hunts other animals for food.

Index

appearance, 4, 12

care, 16, 19–20

enclosure, 16, 19

food, 7, 12, 21

habitat, 8

habits, 4, 10–11, 13, 14, 19–20

tail, 12

Online Resources

popbooksonline.com

Thanks for reading this Cody Koala book!

This book is filled with videos, puzzles, games, and more! Scan the QR codes* while you read, or visit the website below to make this book pop.

popbooksonline.com/l-gecko

*Scanning QR codes requires a web-enabled smart device with a QR code reader app and a camera.